THE HOW AND WHY ACTIVITY WONDER BOOK OF
PREHISTORIC ANIMALS

Written by Helene Chirinian
Illustrated by Kaye Quinn

Endorsed by:
Sue Gruber, B.A. Geology
Elementary Teacher
Santa Rosa, California

Copyright © 1989 by RGA Publishing Group, Inc.
Published by Price Stern Sloan, Inc.
360 North La Cienega Boulevard, Los Angeles, California 90048

Printed in the United States of America. All rights reserved.
No part of this publication may be reproduced, stored in a retrieval system or transmitted, in any form or by any means, electronic, mechanical, photocopying, recording or otherwise, without the prior written permission of the publisher.

10 9 8 7 6 5 4 3 2 1   ISBN: 0-8431-4297-9

PRICE STERN SLOAN
Los Angeles

Introduction

When dinosaurs ruled the Earth, they and other reptiles were not the only animals on our planet. Insects, fish and amphibians were also present. But, more important to us, small mammals had already appeared. The dinosaurs probably didn't even notice these small creatures, but may have stepped on more than a few of them. After about 140 million years the dinosaurs began to disappear. We still don't really know why these giant reptiles became extinct, and we may never discover why the dinosaurs disappeared. We do know, however, that the time following dinosaurs became the Age of Mammals.

The climate in which dinosaurs lived had changed. As this happened, the land was changing, too. Swampy lowlands changed to grassy highlands. Then, animals as fantastic as the giant dinosaurs that came before them appeared: giant mammals! Some of these giants were twice the size of today's elephants. Mammals have changed greatly over the millions of years they've existed.

What are mammals and how are they different from other animals? Mammals are warm-blooded animals with backbones and fur or hair. They give birth to babies rather than lay eggs (except for a few Australian animals such as the duck-billed platypus and the echidna) and feed their babies with milk from their bodies.

Look at the pictures on this and the next page. Draw a line from each giant animal to the word you think is its name.

GIANT HORNLESS RHINO
GIANT CAMEL
GIANT LION
GIANT DEER
GIANT GROUND SLOTH
GIANT PIG
SABER-TOOTHED CAT
IMPERIAL MAMMOTH

Early Mammals

Early mammals were probably nocturnal. This meant that they were active during the night and slept during the day. Nocturnal animals exist today, too. Some of them are hamsters, skunks and raccoons.

CYNOGNATHUS

Although the only living egg-laying mammals that exist today are the duck-billed platypus and the echidna, there were probably many more in the days of the first mammals. CYNOGNATHUS (SY-no-NAY-thus) looked more like a mammal than a reptile. It probably had hair, but had reptile-like features as well. This type of animal we call a mammal-like reptile and we don't know if it laid eggs or had babies. Cynognathus lived about 220 million years ago.

TRICONODON (try-CON-o-don) is considered to be one of the very first true mammals. It lived about 190 million years ago.

PURGATORIUS (PURG-a-TOR-ee-us) was another early mammal, and is considered to be the first PRIMATE (PRY-mate), the classification that includes monkeys, apes and humans. This mammal lived about seventy million years ago, probably laid eggs and was the size of a rat. What happened to these early mammals? Many of them died out because they couldn't change with their world. Others changed over millions of years and their descendants became the animals that we know of today.

Below are pictures of some prehistoric animals. Color them to match the animals you think they became.

Types of Mammals

PLACENTAL (pla-SEN-tal) MAMMALS are the most numerous on Earth. They are warm-blooded and develop in their mother's womb. There, they get food directly from the mother from a special part called the placenta. These babies are born alive and begin to drink their mother's milk immediately after birth. Placental animals include cats, dogs, horses, cows, bears and humans.

MARSUPIAL (mar-SOOP-ee-al) MAMMALS live only in Australia, North America and South America. Marsupials are born alive too, but are very, very tiny at birth. As soon as they are born they crawl to their mother's pouch and complete their development there. They don't leave the pouch permanently until they are able to take care of themselves. The most famous marsupials are Australia's kangaroos, wallabies and koalas. The only marsupials in America are opossums.

MONOTREME (MON-o-treem) MAMMALS only live in Australia, and they are the strangest animals in the world. The duck-billed platypus and the echidna (ek-ID-na) are monotremes. These animals seem to be leftovers from the times of the earliest mammals, many millions of years ago. Monotremes lay eggs, yet have fur and give milk to their young. Their milk glands, however, are different from those of other mammals.

Unscramble the names of the following animals. Write placental, marsupial or monotreme next to each unscrambled name.

musopso _____ _____

gorknaoa _____ _____

dinache _____ _____

nuham _____ _____

aloka _____ _____

owc _____ _____

supyatlp _____ _____

odg _____ _____

reveba _____ _____

The Eocene Period

Fifty million years ago, the dinosaurs had already become extinct. This was called the EOCENE (EE-o-seen) PERIOD. Mammals became the kings of the Earth. Like animals today, some flew in the air, some lived on land and others took to the water. One of those which made the water its home was BASILOSAURUS (BA-sil-o-SORE-us), a primitive whale. About seventy feet long, its legs had become flippers and it had sharp teeth.

Crocodiles and turtles were reptiles that developed during this period. They swam in swamps, not in places like South America, but in southern England. That country's climate was not as cool and damp as it is today.

Many of the animals that lived during this period are not the ancestors of animals living in our time. For many reasons, they died out. Perhaps they were too slow, too clumsy, or unable to change as the Earth itself changed. UINTATHERIUM (YOU-in-ta-THEER-ee-um) was one of these. It was the biggest land mammal of its time, a plant-eater that ate the leaves off trees.

Bats began to fly during this period. An unexpected enemy of mammals was not another mammal, but a seven-foot-tall bird called DIATRYMA (DY-a-try-ma), which stalked small mammals such as HYRACOTHERIUM (HY-ra-co-THEER-ee-um), the ancestor of the horse. Diatryma was almost as tall as an African elephant. It could not fly and became extinct about forty-five million years ago.

Make as many words as you can from these words:

Unitatherium	Diatryma	Hyracotherium

UNITATHERIUM

HYRACOTHERIUM

DIATRYMA

The Oligocene Period

The animals of the OLIGOCENE (ole-IG-o-seen) PERIOD of thirty-eight to twenty-six million years ago were a lot more like the animals of today. Cats, dogs, hyenas, horses and rhinos appeared. Yet, there were still some animals that were doomed to extinction. The most spectacular of these was INDRICOTHERIUM (IN-dric-o-THEER-ee-um), more than twice the size of a modern elephant, at eighteen feet tall and twenty-five feet long. This giant was probably the largest land mammal that ever lived. It probably ate high vegetation as the giraffe does today. Indricotherium was not the only giant of this age. ARSINOITHERIUM (AR-sin-oy-THEER-ee-um) and BRONTOTHERIUM (BRON-to-THEER-ee-um) were absolutely huge! Both these mammals were descended from the same group of animals as horses. This was the period in which hoofed animals appeared. They lived in Africa on the grasslands, had horns and looked somewhat like the modern rhinoceros.

Work the crossword puzzle below.

Across

3. Probably the largest land animal that ever lived
5. Animals of the Oligocene Period that exist today
7. Brontotherium is the ancestor of the same group as these
8. A modern animal that eats high vegetation
10. Indricotherium did not live in water, but on _____

Down

1. Another animal of today that appeared during the Oligocene Period
2. Indricotherium probably ate _____
4. Indricotherium was probably twice the size of a modern _____
6. Early hoofed animals probably lived on the continent that is now called _____
9. Many animals of this period became _____

The Miocene Period

The MIOCENE (MY-o-seen) PERIOD lasted from twenty-six to seven million years ago. During this time, grasses became very important plants. Earth's climate became much drier. As the plants changed, the animals had to, also. Hoofed animals became very numerous, and included both even-toed animals such as the deer, and odd-toed animals such as the horse. MASTODONS (MASS-to-dons), huge hairy ancestors of the elephant, thundered across plains and AEPYCAMELUS (EP-ee-cam-EEL-us), an ancestor of the camel, grazed on the prairies of North America. SABER-TOOTHED CATS, with their long, deadly teeth, preyed upon mammals much larger than themselves.

During this time, animals appeared that mystify people today. One of these animals was MOROPUS (mor-O-pus), an animal whose fossils have been put together in many different ways. This relative of the horse and rhino looks like Dr. Frankenstein put it together! With a horse-like head and teeth, claws instead of hooves and a thick body and legs, the secret of this "mystery mammal" may never be revealed, but some scientists think it may have used its claws to pull down branches and dig up roots and tubers.

Another mystery mammal was MACRAUCHENIA (MAC-ro-CHEEN-ee-a). This South American mammal may have had a trunk, a nose like a camel or even nostrils high up so it could submerge in water and still breathe!

Put the mystery mammal together. Copy each part of the puzzle in the correct square.

The Pliocene Period

By the PLIOCENE (PLY-o-seen) PERIOD, from seven to two million years ago, Earth's animals looked pretty much the way they  do today. Elephants, deer, pigs, giraffes and many types of antelopes lived on the continents of Europe, Asia and Africa. GLYPTODONTS (GLIP-to-donts), the huge ancestor of the modern armadillo, roamed South America. This armored mammal was ten feet long, and early versions of it had sharp spikes on the tail. Both the armor and the spikes protected this armadillo ancestor from fierce predators such as THYLACOSMILUS (THY-la-co-SMILE-us), a cat-like marsupial.

Where do humans, the most advanced mammals on Earth, come in? We do know, of course, that humans didn't yet exist at the same time as any dinosaurs. However, during the Pliocene period the ancestors of modern humans were already beginning to evolve. Placental mammals, they would soon rule the Earth because of their high intelligence. But at this point, they were still  quite ape-like. But by the end of this period, AUSTRALOPITHECUS (OZ-tray-lo-PITH-ek-us), a pre-human, began the evolution from ape to human. It walked upright, lived on the ground instead of in trees, and may even have used stones as weapons.

Write the numbers 1, 2, 3, 4 to show the order in which you think humans developed.

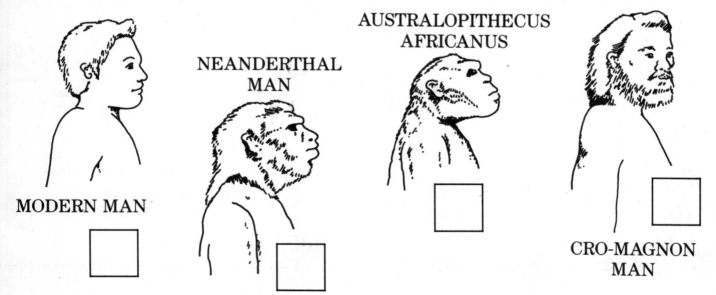

Charles Darwin and Evolution

In 1831, Charles Darwin sailed around the world in a ship called the HMS Beagle. The mission of the Beagle was to map the coasts of South America, but Darwin went along as the ship's NATURALIST, someone who studies nature especially by observing plants and animals. When Darwin began his voyage, like everyone else in his day he did not doubt the idea that animals did not change, but were the same in the past and would be the same in the future as they were in the present. This idea was called THE IMMUTABILITY OF SPECIES. As he traveled, Darwin wondered why similar animals such as the South American rhea and the African ostrich lived so far apart. He also wondered why places very close to each other had animals that were similar but not exactly the same. When he traveled to the Galapagos (Ga-LOP-a-gose) Islands, off the west coast of Ecuador, he found

---

Trace the animals that still exist. Color the animals that no longer exist.

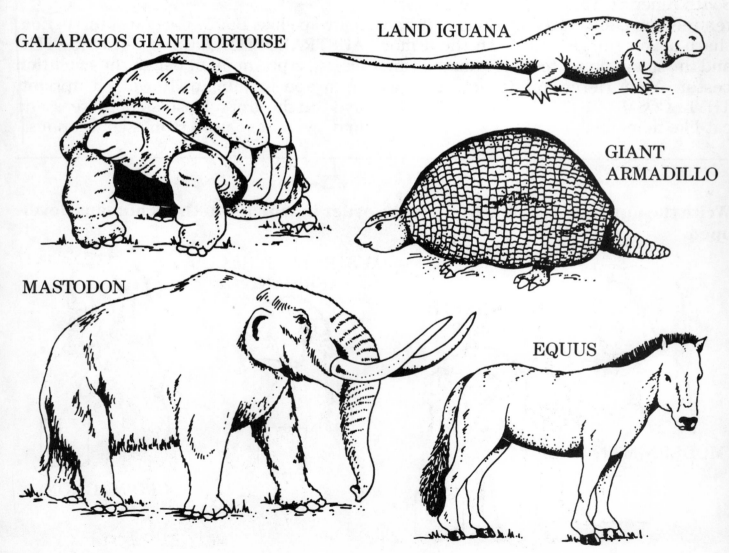

many animals that were similar to, yet slightly different from, animals on the mainland. He even found differences in the same kinds of animals on different islands. He found similar things when he looked at the fossil record. Darwin felt that the only process to explain this was EVOLUTION, the slow changing of plants and animals. To convince others, he needed to know how evolution worked. He knew that animals of the same species showed differences and that some of these differences might make some animals better adapted to their places in nature. He then reasoned that those that were better adapted to the environment or changes in the environment would survive and reproduce and those that would not, would die. This explanation of the way evolution works is called NATURAL SELECTION.

Mammals Evolve from Reptiles

Reptiles didn't change into mammals overnight, and all reptiles didn't become mammals. But many prehistoric animals evolved from reptile to mammal very slowly, over millions of years. Some animal fossils show both reptile and mammal features. DIMETRODON (dim-EE-tro-don) looks enough like a dinosaur to confuse anyone who isn't a scientist, but this animal had a mammal's teeth and a reptile's scaly skin. Most interesting of all about this creature, however, is the sail on its back. Scientists are not certain about the use of the sail, but experiments show it would have been very good for warming up quickly and keeping these animals warm in cold weather. The sail might have been the first kind of solar heat!

What makes a mammal's teeth different from that of a reptile? Reptiles have only one kind of teeth. Mammals have three kinds of teeth. The lower jaw of a reptile has several bones in it. The lower jaw of a mammal is just one bone.

DIMETRODON

Which jaws are found in mammals? Color them green. Which jaws are found in reptiles? Color them brown.

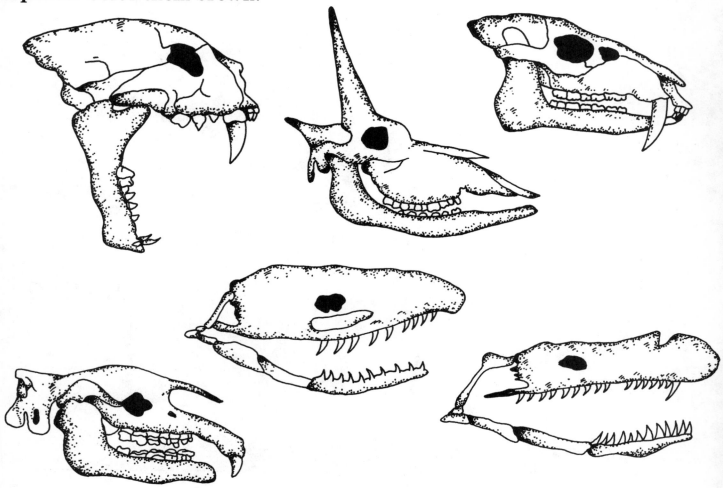

All About Horses

HYRACOTHERIUM (HY-ra-co-THEER-ee-um), the earliest horse, looked nothing like the horses of today. These furry little mammals were about the size of foxes. Hooves had not developed in mammals yet, and these small animals had toenails. Hyracotherium lived about fifty million years ago. Horses grew larger over millions of years. MESOHIPPUS (mez-o-HIP-us) was almost twice as tall as its ancestor. It lived in grasslands and lived about thirty-five million years ago. Ten million years ago, MERYCHIPPUS (mer-ik-HIP-us) roamed the plains. This horse ancestor had already developed hard hooves which helped it run fast. Then, three million years ago, the modern horse, known as EQUUS (EE-kwus), appeared. As the horse evolved, its teeth became larger and sharper, to enable it to eat the changing plant life, as its food changed from soft tree leaves to hard grass stems.

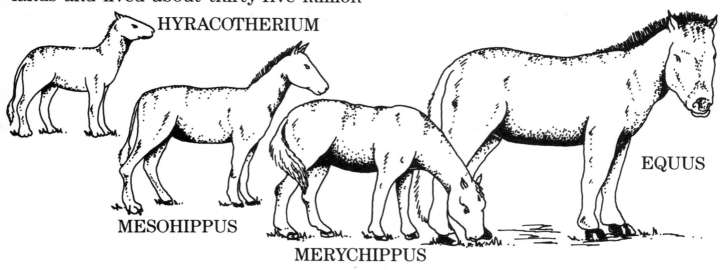

Draw this horse ancestor. Follow the pictures square by square on the empty squares.

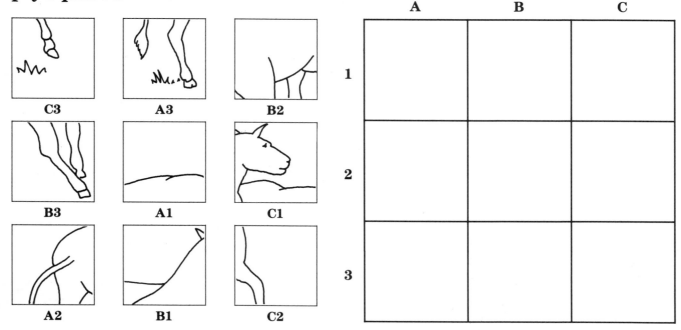

13

All About Elephants

The elephant is the largest living land animal, although forty million years ago its ancestor was about the size of a large pig. This ancestor, called MOERITHERIUM (more-i-THEER-ee-um), had neither trunk nor tusk. It did have a long snout and a tooth that stuck out of either side of its mouth. This animal lived in swamps and ate soft, juicy plants. It protected itself by hiding in the water when danger was near.

Moeritherium's eyes and ears were placed high on its head so it could see and hear when it was hiding underwater.

Elephants got larger and larger. Their upper lips and noses became very long and developed into trunks. Trunks were found to be handy for feeding and drinking. Why didn't the elephant develop a long neck like the giraffe? Its head was much too heavy to be supported by a long neck.

Work the puzzle to see what you know about early elephants.

Across

3. The largest land animal
5. Moeritherium could see when it was _____
6. Moeritherium's ears and _____ were placed high on its head
7. Moeritherium was about the size of a large _____
8. What elephants' noses and upper lips developed into

Down

1. The elephant doesn't have a long neck because its head is too _____
2. Moeritherium had a long _____
4. This early elephant lived in _____
9. The giraffe has a long _____

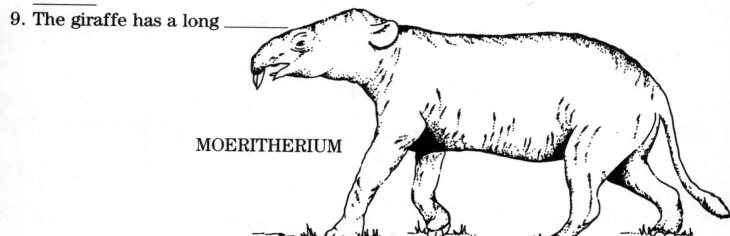

MOERITHERIUM

DEINOTHERIUM (DY-no-THEER-ee-um) lived about fifteen million years ago. It was much larger than the elephants of today. Its tusks were quite different, too, and curved under the lips toward the chest. It looked a lot like the modern elephant. GOMPHOTHERIUM (GOM-fo-THEER-ee-um) was an elephant ancestor that became extinct. This mammal had tusks in both its upper and lower jaws. It lived only six million years ago, but did not evolve into the animal we know today. PLATYBELODON (PLAT-ee-BEL-o-don) looks like a cross between an elephant and a hippopotamus. It had two short tusks in its upper jaw and huge, square tusks in its lower jaw.

Number the elephant ancestors 1, 2, 3, 4, to show the order in which they lived.

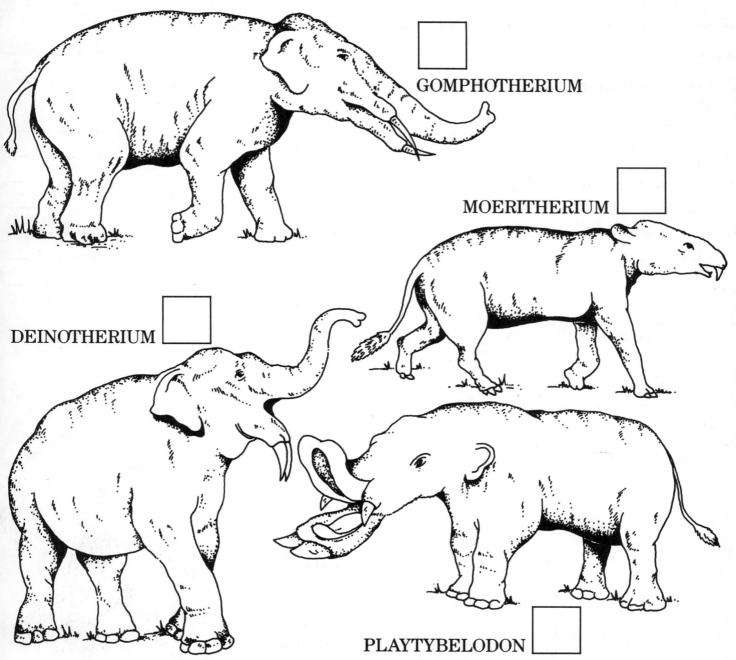

GOMPHOTHERIUM

MOERITHERIUM

DEINOTHERIUM

PLAYTYBELODON

The Biggest Mammals That Ever Lived

Imagine a mammal eighteen feet high at the shoulder, more than twenty-six feet long and weighing eighteen tons. When this animal stretched, the end of its nose could have reached the roof of a one-story house. You don't have to imagine this mammal anymore — it really existed, was an ancestor of the rhinoceros and is called INDRICOTHERIUM (IN-dric-o-THEER-ee-um). It lived thirty million years ago. Much later rhinos, such as DICERORHINUS (DY-sa-ror-HY-nus), lived more than a million years ago. Their size had shrunk until they were not even as large as modern rhinos. ARSINOITHERIUM (AR-sin-oy-THEER-ee-um), an early ancestor that lived thirty-five million years ago, had horns made of bone. Fossils of this slow, heavy animal's horn have been found, but the horns of later rhinos were not preserved well as fossils. This is because the horns of later rhinos, like the rhino of today, are made of bunches of hair stuck together, not bone! However, we can tell which rhinos had horns because of lumps on the skull where the horns grew.

Make as many words as you can out of ARSINOITHERIUM.

_____   _____   _____
_____   _____   _____
_____   _____   _____
_____   _____   _____
_____   _____   _____
_____   _____   _____
_____   _____   _____

BALUCHITHERIUM

ARSINOITHERIUM

Early Cats

Cats have always been very good hunters. Since cats first appeared, they have hunted by stalking their prey, then pouncing on it. With claws that could hold prey while killing it, early members of the cat family were fearsome predators.

SABER-TOOTHED CAT

The SABER-TOOTHED CAT is one of the most well-known prehistoric mammals and lived about twenty-six million years ago. There were many different kinds of saber-toothed cats, whose fossils have been found at the La Brea Tar Pits in Los Angeles, California. Their long, knife-like teeth were most likely used to tear elephant hide, which is very thick and tough. Most cats have RETRACTABLE CLAWS, which means they pull their claws back into the skin of their paws so they won't become dull. And just to make sure their claws are always ready to strike, cats have always sharpened their claws on tree trunks. Even the small house cat we keep as a pet today likes to sharpen its claws on furniture or special pieces of wood given to it by its owner.

Help the elephant escape from the saber-toothed cat.

How Dogs Evolved

The dog was the first animal domesticated by humans. This means that humans tamed the dog and kept it to do work and as a companion. We all know that dogs evolved from wolves into the hundreds of different types available as pets today, mostly through selective breeding by people. However, there are still wild dogs that hunt in packs in Africa, and dogs that are not kept as pets will quickly go back to a wild state and learn again to hunt in packs. Wolves still exist, and some humans are still trying to tame them, not satisfied with the dogs they have bred. PSEUDOCYNODICTIS (SOO-do-CY-no-DIK-tus) lived about thirty million years ago. They were about the size of a modern weasel, lived in packs and hunted together.

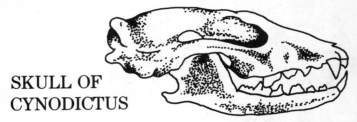

SKULL OF CYNODICTUS

Change the word from WILD to TAME in five moves. Change only one letter with each move.

WILD
TAME

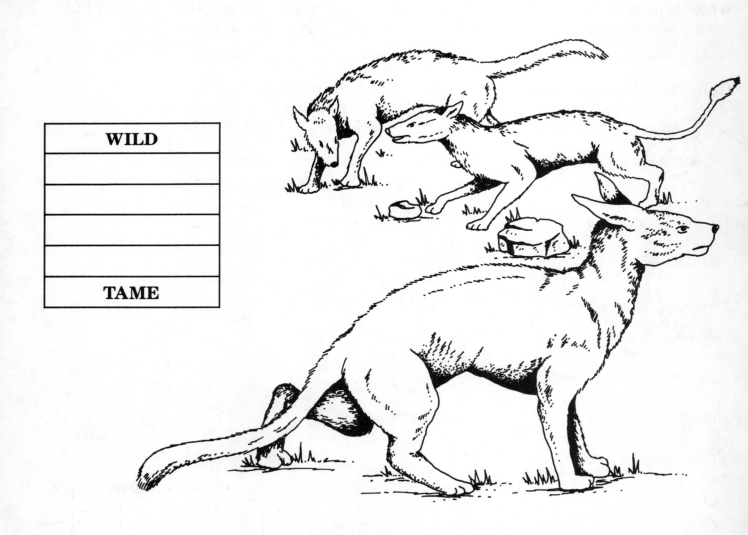

Surprising Ancestors

Suppose you were visiting a river, saw a beaver dam and went up to take a closer look. But, what if, instead of a cute, small creature we know as a beaver, a 500-pound giant lumbered out of the dam? Does that

STENOMYLUS

sound like a science-fiction movie? Guess again — beaver ancestors did weigh one-quarter of a ton! Early horses were not much bigger than house cats. If they were alive today, you could pick them up and carry them around! An early camel, called STENOMYLUS (STEN-o-MY-lus), was the size of a modern goat. Perhaps it had a hump, perhaps not. And today's elephants have very little hair. Imagine seeing a WOOLLY MAMMOTH, its Ice-Age ancestor, with a coat as furry as that of a bear! And, a GIANT SLOTH, as large as an elephant, would make its relatives that live today look like dwarfs!

WOOLLY MAMMOTH

Write the answers. Then write each letter in the box with the same number below. The answer is the name of an animal hunted by our ancestors.

1. A giant sloth of long ago would make its relatives today look like __ __ __ __ __ __ .
      1 8 2 3 11 13

2. If you saw a beaver dam you might take a closer __ __ __ __ .
     11 5 5 3

3. Ancient sloths were as big as a modern __ __ __ __ __ __ __ __ .
     15 6 15 20 14 2 3 26

4. __ __ __ __ __ horses were not much bigger than today's house cats.
    15 2 3 6 4

5. Beavers build __ __ __ __ .
                1 2 12 13

6. Stenomylus was an early __ __ __ __ __ .
                            2 9 12 15 23

7. This early camel was the size of a __ __ __ __ __ __   __ __ __ __ .
                                    12 5 1 15 3 24   19 5 2 26

8. Stenomylus may not have had a __ __ __ __ .
                               10 17 3 2

8	5	5	6	6	4	12	9	12	12	5	26	10

How Fossils Are Made

Rancho La Brea in Los Angeles, California is a goldmine of prehistoric mammal fossils. The reason for this is that this part of the busy city of Los Angeles is an ancient tar pit. The La Brea Tar Pits, as they are known, contained a thin layer of rainwater on the surface and sticky tar beneath. Animals would wade in to drink and become stuck in the tar. The tar was so sticky that the animals could not escape and get back on the shore. As the frightened animals struggled to get out of the tar, predators saw them as easy pickings that couldn't get away. As the wolves and saber-toothed cats pounced upon the mammoths (a type of elephant) and other trapped animals, they found themselves in the same trouble, trapped on an animal that couldn't escape from the tar pits. The tar hardened slowly and the bones of the trapped mammals became preserved as fossils. The tar pits are fenced off, but small animals still fall into them. Mice, gophers, squirrels, birds and insects from today may be found by people thousands of years from now.

Follow the correct path to the animal that got stuck in the tar pits.

Another way fossils are made is by bones being buried for thousands, or even millions, of years. When a prehistoric animal

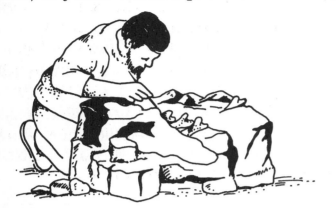

died, it might be covered by dirt. Rain or bodies of water would turn the dirt into mud. The mud would become hard as rock. More and more mud and rock would pile up. The bones would be the only part of the animal left, because the softer parts would rot away and disappear. Bones can be dug up unchanged, or they can be dug up as fossils that have turned into minerals. Some bones, although hard, dissolve in time. They leave a hole, like a mold, in the rock. Footprints also become fossils. Teeth, which are even harder than bones, are found very often. But, perhaps the most interesting kind of fossils are those from the Ice Age that have been found very well preserved in the ice. Both woolly rhinos and woolly mammoths have been found. From these fossils, scientists have been able to discover the color and shape of these animals, and even what they ate at their last meals!

The words on the list below are hidden in the sentences. Find and circle them.
EXAMPLE: John wore a suit o**f arm**or.   **FARM.**

        bones   **rain**   **water**   **often**

1. Digging up fossils was his job one summer.
2. We saw Laura in the mud searching for woolly rhinos.
3. In Ottawa terrific museums are found.
4. We played a game of tennis after the picnic.

WOOLLY MAMMOTH   WOOLLY RHINOCEROS

The Ice Age

Ice Ages come and go. During the last million years, the northernmost and southernmost parts of the world have been covered with thick ice. There was a short spring or summer and the snow and ice never melted. Many animals couldn't stand the cold and moved where it was warmer. Other animals stayed in the cold, but to do so, they had to evolve for survival. Reindeer, bison, wolves and some hares still live in cold-winter places such as Canada, the northern United States and Scandinavia. They also lived during the Ice Ages. But, there were surprises. Elephants, which now live only in Africa, India and some other Asian areas lived in the frozen north. Rhinos, now found in Africa, were also survivors of the Ice Ages. But to survive, these animals had to grow thick coats of hair and a thick layer of fat to keep warm. WOOLLY MAMMOTHS also had long, curved tusks, which were probably used like snow plows to uncover plants to eat. WOOLLY RHINOS had two horns for protection.

Follow the lines to find an animal that lives in a cold place.

Prehistoric Animals and Humans

Although human ancestors missed both the age of dinosaurs and the early age of mammals, they were around to see animals such as the woolly rhino and woolly mammoth. The mammoths were a valued source of food, shelter and clothing for people. The meat was eaten and the hides were used for clothing and stretched over poles to make shelter. Human ancestors were also in great danger from wolves, cave lions and cave bears, which preyed upon them. At the same time, they began to domesticate wolves and other animals. Cave-dwellers painted pictures of many animals on the walls of their caves. Through these cave paintings, the most famous of which are in Spain and France, we learn about the size and colors of cave lions and bison, and we see scenes of our ancestors hunting.

Find the ten hidden pictures that show what you might find in a cave during one of the Ice Ages.

Some Change, Others Don't

Evolution doesn't run smoothly and it's hard to know how an animal will evolve. For example, many animals have hardly changed over millions of years. An example of one of these is the opossum, which has looked pretty much the same for sixty million years. Bats haven't changed much in fifty million years. Why not? Maybe they were just lucky and found a place on Earth that another animal would not take. Maybe they evolved so well the first time that little change was necessary. But some mammals changed greatly. Even-toed, hoofed mammals have become a very large group that include sheep, deer, pigs, hippos, camels and a huge number of antelopes. Why has this group done so well? We don't really know the answer. Human beings have evolved from mammals that lived in trees, could make only a few sounds and knew how to use only one or two tools. We have become people who can live in tall buildings, speak many languages and have invented many machines to work for us. Has evolution stopped? Probably not. It would be fun to see Earth millions of years from now to know what mammals, including humans, look like!

Look at the pictures below. Finish the time line by drawing a picture of what you think the person of the future will look like.

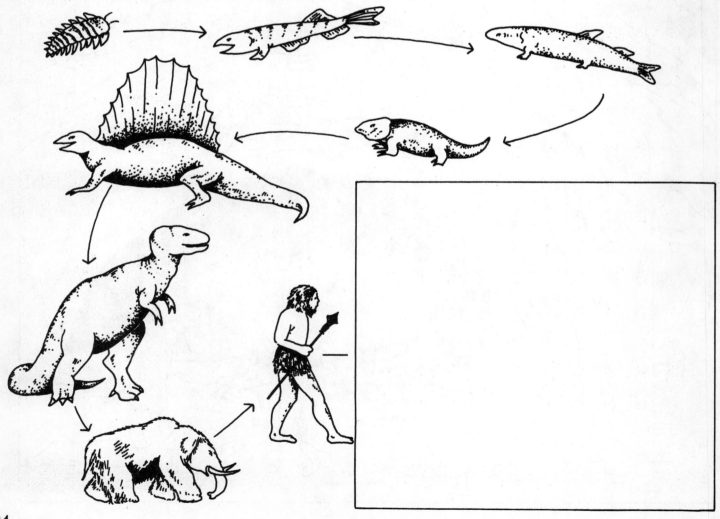

24

The Ancestor of All Carnivores

A CARNIVORE (CAR-niv-ore) is a meat-eater. Many animals, such as the cat, lion and wolf, are carnivores. Others, such as the squirrel, horse and cow, are HERBIVORES (HERB-iv-ores), which means they eat vegetables. Humans are OMNIVORES (OM-niv-ores), which means we eat both meat and vegetables. Our early ancestors, before they learned to choose non-poisonous plants to eat or learned how to farm, were carnivores. Carnivores do eat other carnivores, but many prefer to eat herbivores. MIACIS (mee-AH-sis) is called by scientists "the ancestor of modern carnivores." This ancient animal lived about fifty million years ago, and looked like a modern weasel, a cat and a dog! It also had a tail like that of a raccoon, yet, it is not any one of these animals. A complete skeleton of this strange animal has never been found, but scientists think it spent a lot of its time in the trees.

Find and circle the words from this list that are hidden in the word search below. The words are hidden vertically, horizontally and diagonally.

CARNIVORE   RACCOON
HERBIVORE   ANCIENT
OMNIVORE   WEASEL
MIACIS   TAIL
PLANTS   SKELETON

```
H T G E P B X S R C
E G T Q H D E K L A
R L Q A T X L E E R
O C M U I J R L S N
V S T N A L P E A I
I A N C I E N T E V
N S I C A I M O W O
M R A C C O O N F R
O H E R B I V O R E
```

A Bear-Dog

It's hard to believe that bears and dogs had the same ancestor, but to understand evolution is to understand that all mammals came from the same ancestor, we just went down different roads. Early ancestors looked very different from mammals of today, but had many of the same traits, too. This animal, DAPHOENUS (daf-EE-nus), which lived in the American West about twenty-five million years ago, had a skull much like a dog's, feet, hind legs and lower backbones like a cat's and had eating habits like those of both a bear and a dog. What will the animals of the future look like? People have already bred sheep and goats, to get "geeps" and tigers and lions to get "tigons." Horses and zebras can mate, and horses and donkeys have mated for a long, long time, producing mules. Mules, however, cannot breed. Wild dogs and coyotes produce young called "coys." Evolution, sometimes with the help of humans, continues.

DAPHOENUS

Look at the animals below. Draw lines to match the pairs that you think would make good "animals of the future."

CAMEL SQUIRREL MOOSE DEER LLAMA BEAVER

Name your new animals _____ _____ _____

The First Cattle

Cows and other cattle are not considered to be very smart. They move slowly, moo a lot and stand around while being milked, except for bulls, of course. However, cattle are very important to the civilization of humans. Cattle now extinct were all over Earth ten million years ago. Fossils of very small cattle have been found. Paintings of cattle have been found in caves, showing that early humans kept cattle and probably used them for meat, hides and milk, exactly as they are used today. Today's cattle have shorter legs and horns than prehistoric cattle, but they have many things in common such as hollow horns that are not shed. Early humans captured wild cattle and brought them home to be tamed. They probably tried to capture smaller animals thinking that they would be easier to control. Soon they had herds of cattle. Yaks and buffalos of Asia and Africa are the closest animals to wild cattle that exist today.

Many animals have horns. Find and color the cattle in this picture.

On the Road to Extinction

Many mammals that lived in the past never made it to the present. Some died out many millions of years ago. Maybe they were just like experiments that didn't work. Maybe they were too big, too small, too heavy, needed too much food or were too clumsy to get away from predators. But many mammals have become extinct recently, and many more are on the road to extinction. When Darwin wrote about natural selection he didn't mean people making animals extinct. Humans are the greatest enemies of other mammals. The panda is now very rare because its territory has been reduced. The leopard and other beautiful big cats are in danger because thoughtless people want their fur for coats. The rhino is threatened because some people think its horn has magical powers, so it is simply killed for the horn. Elephant tusks make ivory jewelry and pieces of art that people want to show off, so elephants are killed only for their tusks. Even the mighty gorilla is in danger because the forests that are their homes are being cut down very quickly. As people move into areas that belonged to animals, the animals that can't adapt, die.

Woolly mammoths died out because the Ice Age ended. They couldn't live in the warmer climate. They and woolly rhinos also died because people hunted them. Saber-toothed cats such as SMILODON (SMY-lo-don) became extinct when the mammoths disappeared because mammoths were their main food. Even the huge indricotherium disappeared because the forests disappeared, taking with them the soft leaves this giant fed upon. This animal was too large and too slow to find enough food in the grasslands. Some animals probably will never become extinct. Rats and insects adapt well to many environments. Sharks have been around for millions of years because they are adaptable. How about people? Will we disappear, too? Not if we're smart. If we learn not to pollute our air, soil and water and if we learn to live with other mammals and help them survive, if we don't destroy all our forests and rivers, we will probably be around for a long, long time.

Answers

PAGE 2

PAGE 4

PAGE 5

musopso = opossum, marsupial
owc = cow, placental
gorknaoa = kangaroo, marsupial
supyatlp = platypus, monotreme
dinache = echidna, monotreme
odg = dog, placental
nuham = human, placental
reveba = beaver, placental
aloka = koala, marsupial

PAGE 6

Unitatherium
wait, unite, the, hat, rat, mat, hut, man, hit, matter, nut, near

Diatryma
try, rat, tar, my, day, may, am, mat, dim, ram, dam, at

Hyracotherium
race, the, hat, much, rear, him, eat, hot, moth, each, reach, other

PAGE 7

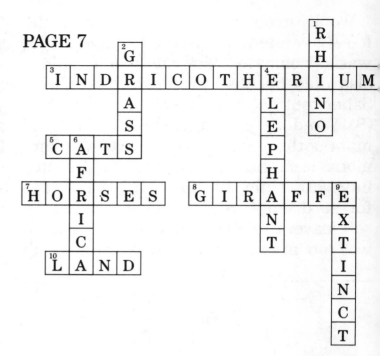

PAGE 8

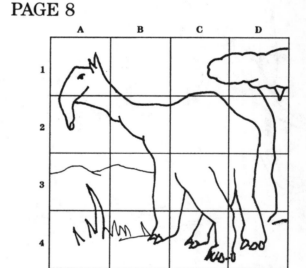

PAGE 9

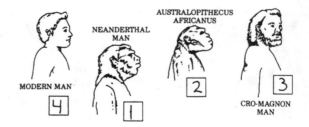

PAGE 10

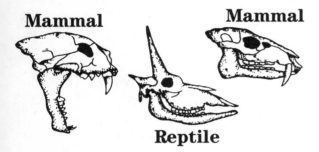

PAGE 12

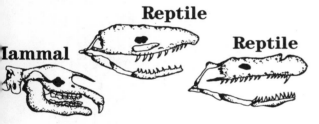

PAGE 13

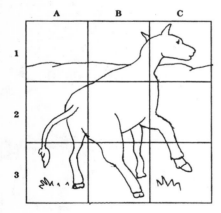

PAGE 14

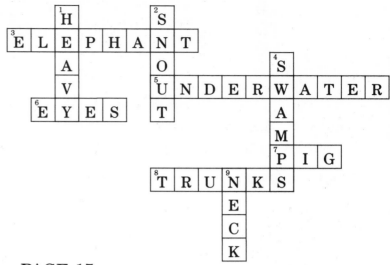

PAGE 15

PAGE 16

Arsinoitherium

the, ear, rest, rinse, east, mother, to, nurse, run, hear, other, it, snore, moth, meat, noise, tear, rat, met, raisin

PAGE 17

PAGE 18   WILD
           MILD
           MILE
           TILE
           TIME
           TAME

PAGE 19

1. dwarfs, 2. look, 3. elephant,
4. early, 5. dams, 6. camel,
7. modern goat, 8. hump

Coded word: WOOLLY MAMMOTH

PAGE 20

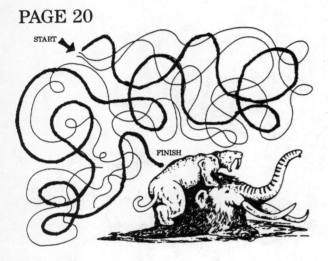

PAGE 21

1. Digging up fossils was his jo**b one** summer.
2. We saw Lau**ra in** the mud searching for woolly rhinos.
3. In Otta**wa ter**rific museums are found.
4. We played a game **of ten**nis after the picnic.

PAGE 22

PAGE 23

PAGE 25

```
H T G E P B X S R C
E G T Q H D E K L A
R O L Q A T X L E S R
O V C M U I J R E A N
V I   S T N A L P   E I
I N   A N C I E N T   V
N     S I C A I M   W O
M     R A C C O O N   R
O H E R B I V O R     E
```

PAGE 27

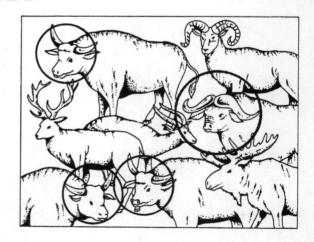